Pollyanna Faria Nogueira
João B. P. Cabral

Spatial-Temporal Analysis Water Quality Framework

Pollyanna Faria Nogueira
João B. P. Cabral

Spatial-Temporal Analysis Water Quality Framework

Surface Waters of the Foz do Rio Claro HPP Reservoir (GO)

Imprint

Any brand names and product names mentioned in this book are subject to trademark, brand or patent protection and are trademarks or registered trademarks of their respective holders. The use of brand names, product names, common names, trade names, product descriptions etc. even without a particular marking in this work is in no way to be construed to mean that such names may be regarded as unrestricted in respect of trademark and brand protection legislation and could thus be used by anyone.

Cover image: www.ingimage.com

This book is a translation from the original published under ISBN 978-613-9-69928-5.

Publisher:
Sciencia Scripts
is a trademark of
Dodo Books Indian Ocean Ltd. and OmniScriptum S.R.L publishing group

120 High Road, East Finchley, London, N2 9ED, United Kingdom
Str. Armeneasca 28/1, office 1, Chisinau MD-2012, Republic of Moldova, Europe
Printed at: see last page
ISBN: 978-620-8-17589-4

Copyright © Pollyanna Faria Nogueira, João B. P. Cabral
Copyright © 2024 Dodo Books Indian Ocean Ltd. and OmniScriptum S.R.L publishing group

Index

Summary

Water resources are invaluable to the environment and indispensable for economic, political and social development. With the growing population there has been an increase in demand for the use of water resources, and this is characterised as one of the **main causes of pollution of water bodies. Another factor is the use of watercourses for** hydroelectric **power generation** between the cities of Jataí and São Simão, changing the physical, chemical and biological characteristics of water resources due to the transformation from a lotic to a lentic environment. Based on this, this study aimed to assess the water quality of the Foz do Rio Claro Hydroelectric Power Plant (HPP) reservoir using the limnological parameters total phosphorus (TP), **chlorophyll "a" (CHL), Dissolved Oxygen (DO),** Turbidity, Total Dissolved Solids (TDS), correlating two distinct periods of the Brazilian cerrado. Based on the standards recommended by CONAMA Resolution 357 of 2005. Twenty-three sampling points were determined in the Foz do Rio Claro HPP lake, due to its size, in order to understand the spatial and temporal distribution.

Keywords: Limnology. Reservoirs. Water quality.

CHAPTER 1

Introduction

According to Rebouças et al (2002), Brazil has one of the largest water reserves in the world, concentrating around 12 per cent of the planet's available surface fresh water, with around 80 per cent of total water production located in the Amazon, São Francisco and Paraná basins.

Even with this large reserve, according to Espindola (2003), the vast majority of river basins in Brazil encompass different regions with different geomorphological and socio-economic characteristics. These regional differences, as well as the diversity of environmental impacts that have occurred and the small database that exists, especially for the majority of aquatic ecosystems. The influence of seasonal and spatial variations is one of the biggest problems for the surrounding communities, especially in terms of the use and management of these systems.

In Brazil, the 1960s and 1970s can be considered the period of major construction of hydraulic projects, mainly for the production of hydroelectricity and human supply (TUNDISI, 2007). Many of these projects are still in operation, producing countless benefits and environmental impacts both locally and regionally.

According to studies carried out by Cabral **et al** (2013) and Rocha, Cabral and Braga (2014), the Claro river basin is heavily used by humans, with urban areas, agriculture (sugar cane, maize and soya) and livestock (cattle breeding), leading to the disposal of domestic and industrial effluents into **watercourses, altering the dynamics of rivers and reservoirs due to the increase in** point source and diffuse pollution **caused by** human activities.

The choice of the Foz do Rio Claro Hydroelectric Power Plant (HPP) as the object of study is justified due to the anthropogenic processes taking place in the Claro River basin, where three HPPs (Caçu, Barra dos Coqueiros, Foz do Rio Claro) and the Irara SHP have been set up **in run-of-river mode. All the projects are located** upstream of the dam, with the Foz do Rio Claro plant being the closest to the mouth of the Paranaíba River.

The Foz do Rio Claro HPP has been in operation since January 2010, its reservoir area is 7.69 km^2 and its energy potential is 64.8 MW (Map 1). Its hydrographic basin is located between the municipalities of São Simão and Caçu, in the state of Goiás, with an area of approximately 151 Km^2 , south of the Itaguaçu district (GO) in the Quirinópolis micro-region (GO).

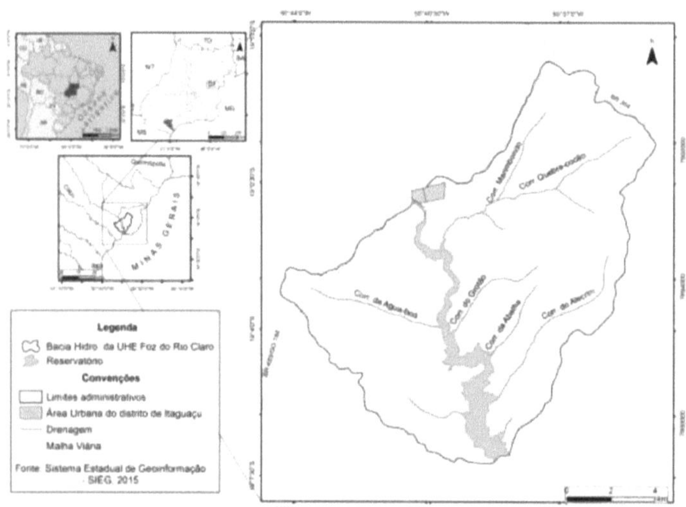

Map 1- Location of the Foz do Rio Claro HPP Hydrographic Basin.

Organisation: Nogueira, P.F. Queiroz Junior, V.S. (2015)

According to data from the Goiás State Geoinformation System (SIEG) (2011), the study area is geologically located in the Paraná

Sedimentary Basin and has two major lithostratigraphic groups of Mesozoic age, formed by the basalt of the Serra Geral Formation of the São Bento Group and the sandstones of the Vale do Rio do Peixe Formation of the Bauru Group (Map 2).

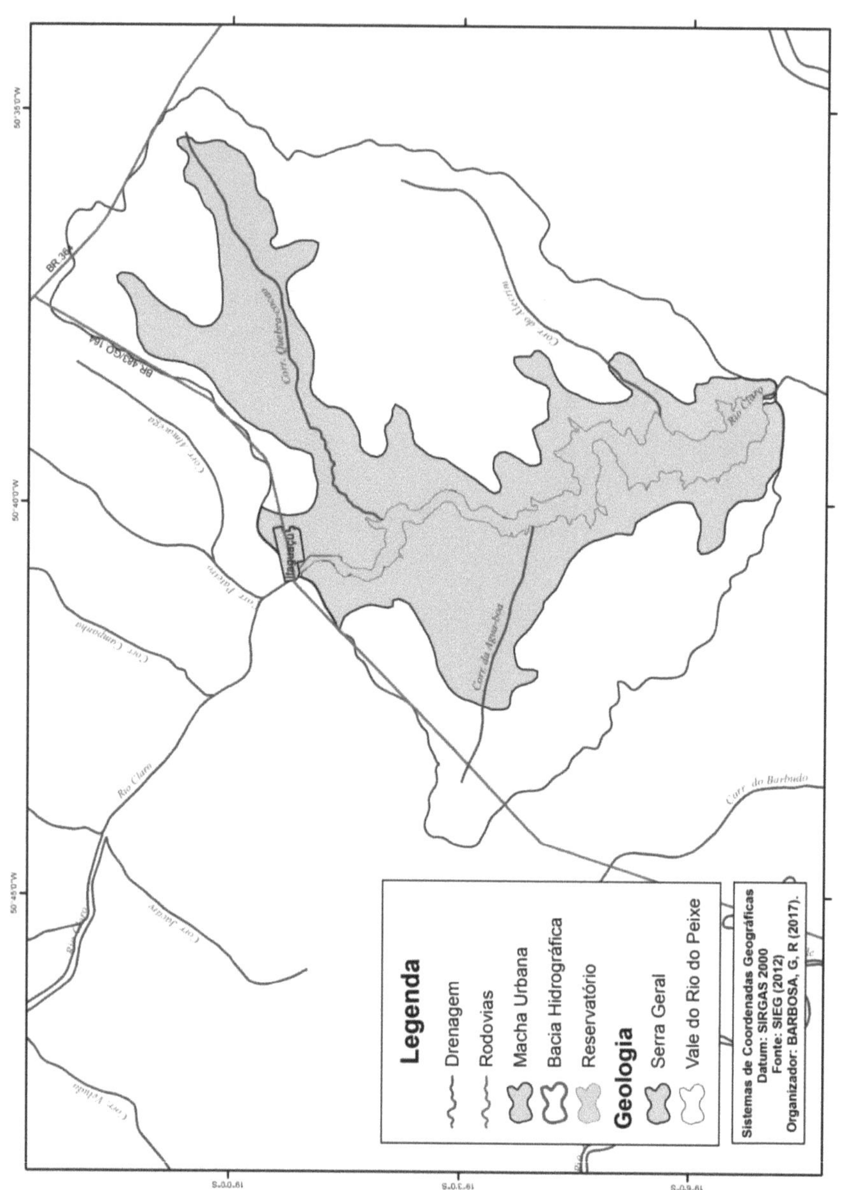

Map 2 - Geology of the Foz do Rio Claro HPP Hydrographic Basin
Organisation: Barbosa, G. R. (2017).

According to Fernandes (1994) and Cabral (2006) the sandstones of the Vale do Rio do Peixe Formation of the Bauru Group are very fine to fine-grained, varying in colour from light brown to pinkish to orange; they are moderately to well sorted. The basalt of the Serra Geral Formation was formed by several episodes of basic lavas that occurred in the Jurassic Cretaceous Radambrasil (1983). The decomposition of the basalt rock (mainly the pyroxene and feldspar minerals) resulted in the genesis of the Red Latosols and their associations with other minerals from other geological formations.According to RADAMBRASIL (1983), the decomposition of the minerals of the basaltic rock (pyroxene and feldspar) of the Serra Geral formation gave rise to the Red Latossolo distroférrico, which is made up of strata of sub-metre thickness, of sandstones interspersed with siltstones or sandy mudstones covered by layers or levels of sandy or sandy-clay sediments.Pedologically in the basin (Map 3), Red Latossolos and Red Latossolo distrófico predominate. The Red Latosols are highly evolved, with significant fertilisation resulting in intense weathering of the primary mineral constituents, with low saturation. The dystrophic red latosol is characteristic of the Bauru Group. It also has similar characteristics to red latosol soils, except that it has a lower supply of iron oxide, with Fe2 and O3 contents equal to or less than 11%, which results in a more yellowish colour.

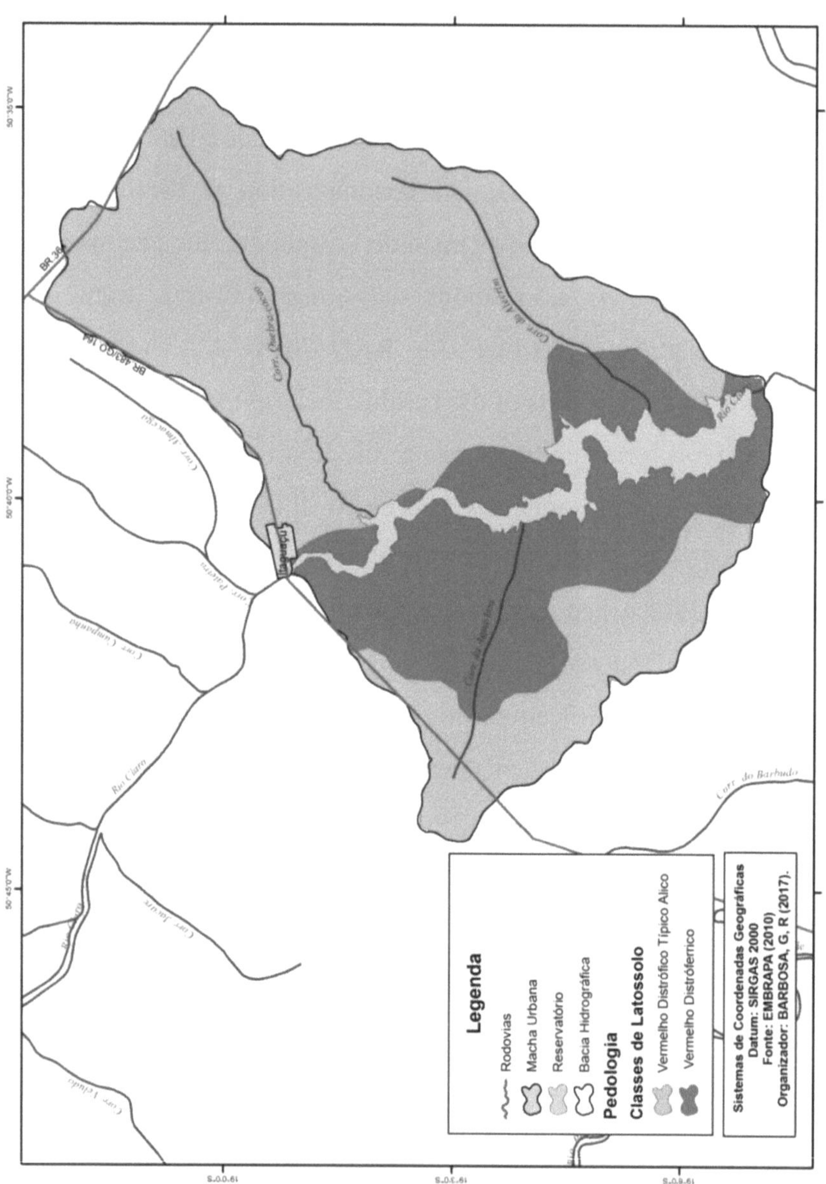

Map 3 - Soils of the Foz do Rio Claro HPP Hydrographic Basin
Organisation: Barbosa, G. R. (2017).

The climate of the region according to Koppen's (1931) proposal and the research carried out by Lima (2013) for the Claro river basin is classified as Awa, tropical savannah, mesothermal. The temporal distribution of the monthly rainfall averages of the four seasons indicates the occurrence of a rainy period that extends from October to April with the highest concentration of rainfall and a less rainy period from May to September. The average temperature of the hottest month is 24 °C, the absolute maximum is around 38 °C; the average temperature of the coldest month is 18 °C and the average annual rainfall is 1750 mm.

According to SIEG (2006), the state of Goiás is characterised by a rainy period (October to April) and a less rainy period (May to September). In the rainy season, 90% of the total rainfall occurs, with the months of December and January standing out, showing that most of the state receives around 250 to 300 mm of rain. The month of December is the wettest, with relative humidity levels of between 80 and 82% in around 50% of the state's area.

According to Queiroz Junior (2014), land use in the study area (map 4) is basically composed of pasture, with a few crop areas, which are sugar cane.

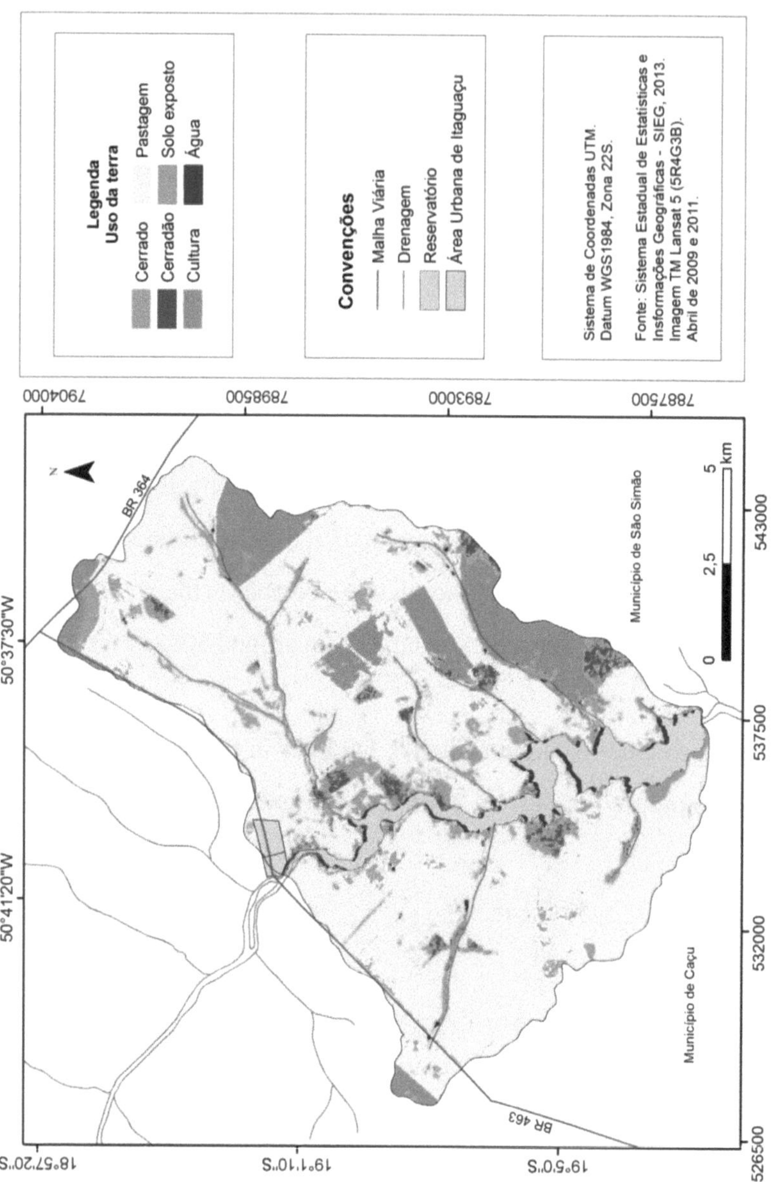

Map 4 - Land use and vegetation cover for 2011 in the Foz do Rio Claro HPP catchment area.
Organisation: QUEIROZ JÚNIOR, V. S. (2013).

According to Queiroz Junior (2014), agriculture increased by 6.63 per cent to a total of 9.54 per cent. He also points out that a large part of the bare soil is used for agriculture, totalling 12.17%. The areas with vegetation fell by 5.21 per cent from 2009 to 2011, where they are mainly occupied by the waters of the reservoir that was built at the time and in small parts by pasture.

Geomorphologically, according to Latrubesse and Carvalho (2005), the region comprises the categories of SRAIIIB-RT(m) - Regional Plateau Surface IIIB with elevations between 650 and 750 m, with medium dissection and associated with Tabular Reliefs in the Paraná Basin to the north-east, another compartment refers to the SRAIVB-LA(fr) Regional Plateau Surface IVB, which has elevations between 400 and 550 m and is considered to be weakly dissected to the south. This compartment is associated with the relief of the Paraná Basin and with lake systems. The relief of the rivers is characteristic of the plateau, which does not favour navigation and is therefore more suitable for the construction of hydroelectric power stations.

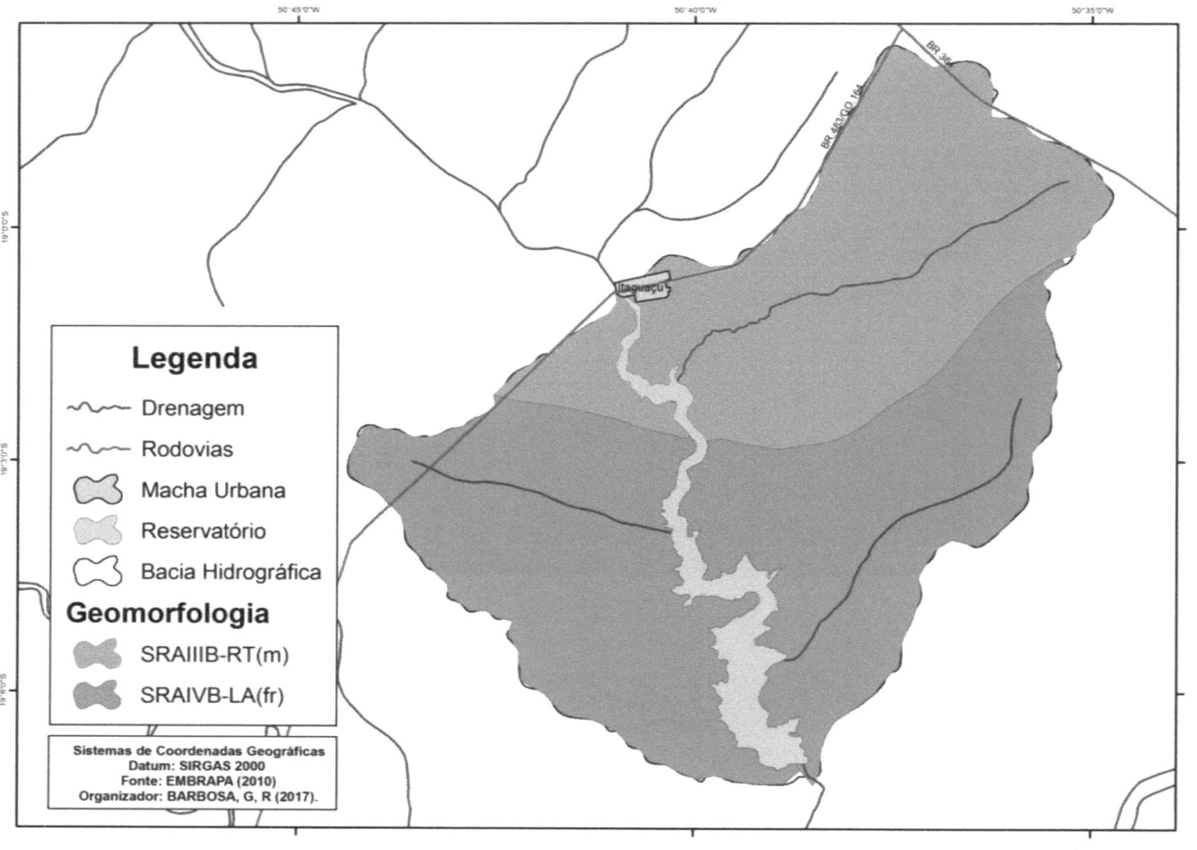

Map - Geomorphology of the Foz do Rio Claro HPP Hydrographic Basin.
Organisation: Barbosa, G. R. (2017)

Twenty-three sampling points were determined in the Foz do Rio Claro HPP reservoir (Map) due to their size, in order to understand the spatial and temporal distribution of the reservoir .

nutrients. The water samples were collected from the first 20 cm of depth, which according to Esteves (1998) and Tundisi (2008) corresponds to the most superficial layer of the body of water (epilimnion) where the highest temperatures tend to occur, which, together with the nutrient input, increases the productivity of organisms in the superficial layers.

At each sampling point, measurements were taken of T, DO, TDS and water samples were collected for the Turb, CHL and PT tests. To determine CHL, we followed the proposal developed by Yunes and Araújo [s/d] based on Mackinney (1941), Paranhos (1996) and Chorus (1999), which is based on calculating the chlorophyll concentration in $\mu g.L^{-1}$ considering the spectrophotometer measurements, the volume of pigments extracted and the filtered volume.

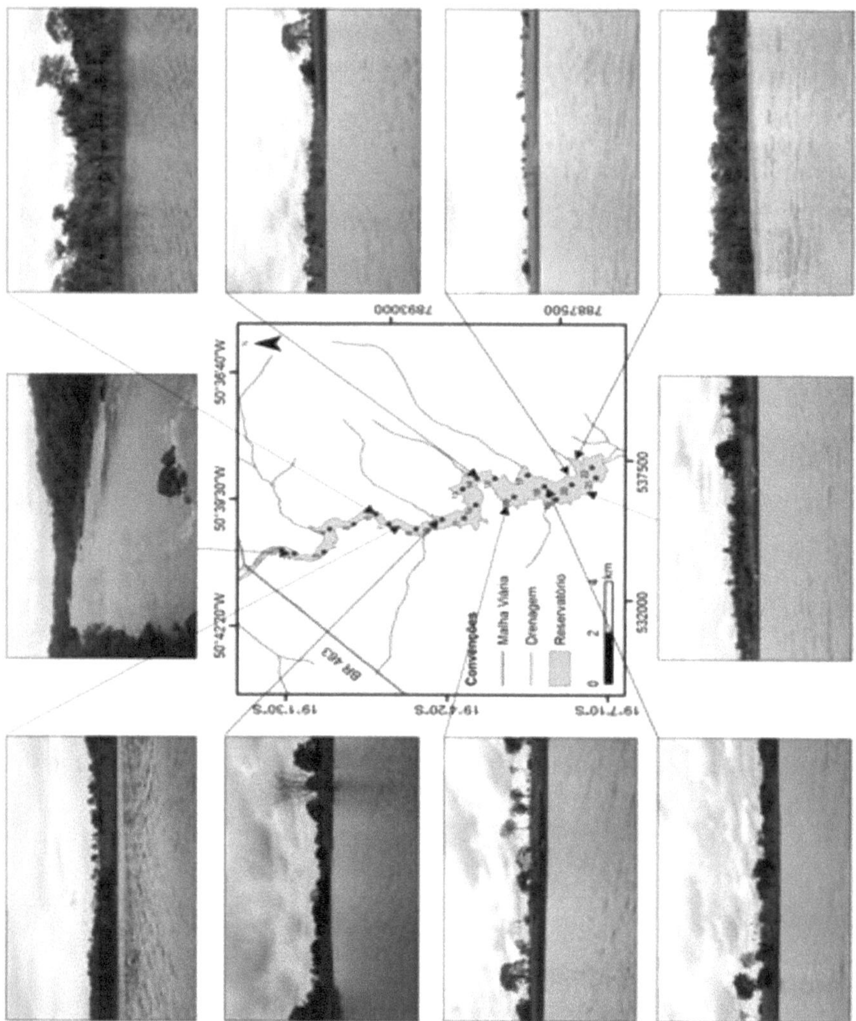

Map 5 - Sampling points and type of land use near the collection points
Source: Nogueira, F. P. Queiroz Júnior, V. S. (2014).

To determine Chlorophyll "a", the proposal developed by Yunes and Araújo [/d] based on Mackinney (1941), Paranhos (1996) and Chorus (1999) was followed. The water samples were cooled until they were filtered.

For the filtering and pigment extraction process, the analyses were always carried out under dim lighting, without the use of lamps and with the environment in **low light, so that there would be no change in Chlorophyll "a".** **Filtering was carried out in the** field and used a 47 mm diameter AP20 Millipore glass fibre filter with a nominal retention capacity of 0.8 to 8.0 microns.

The volume of the water sample was 200 ml, and it was poured in until the filter no longer allowed the water sample to pass through, i.e. once the previously established volume had been reached (filtering should not take more than 15 minutes). The filter It was then removed and placed on absorbent paper to remove the excess moisture.

According to the methodology, after filtering it is recommended to store the samples at a low temperature for a period of three months. The filtering and extraction of the pigments were carried out at the Applied Geosciences Laboratory (LGA), so the filters were wrapped in aluminium foil (folding them in half with the pigments facing inwards) according to Mackinner (1941), Yunese and Araújo [n.d.].

To extract the chlorophyll a pigment, the filters were placed in jars, 10 ml of methanol was added, then the jars were capped and wrapped in aluminium foil and kept refrigerated for 24 hours (photo 1a).

The reading was taken on the spectrophotometer, for which the filters were removed from the refrigerator under dim lighting, and the methanol was pipetted and placed in the spectrophotometer cuvettes with an optical step of 1 cm. According to Mackinner (1941), Yunese and Araújo [s/d], the absorbance of Chl-a occurs at 663 nm and the turbidity of cyanobacterial cells at 750 nm.

To calculate the chlorophyll concentration in $\mu g.L^{-1}$, the spectrophotometer settings, the volume of pigments extracted and the filtered volume are taken into account, as shown in Equation 1. In the experiments carried out, a UV - 1000A Spectrophotometer was used, with an operating range of 320 to 1100 nm from Instrutherm (Photo 1 b).

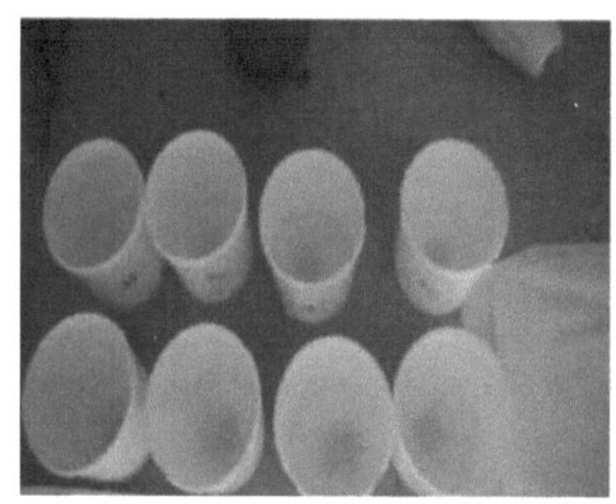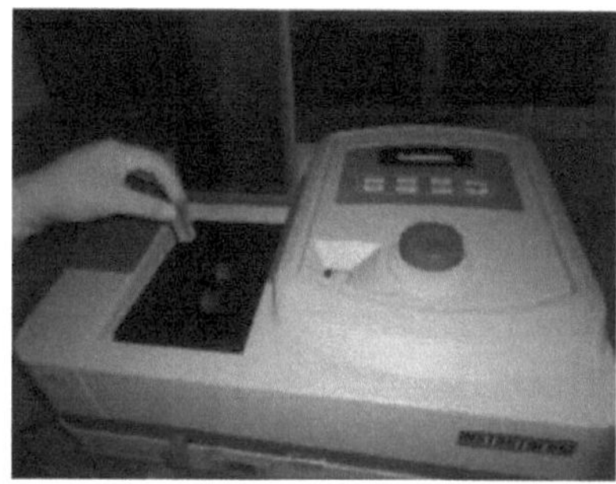

Photo 1 (a) Vials with the filters immersed in methanol.
Photo 1 (b) Spectrophotometer used to read absorbance.
Source: The author himself.

The experiments were carried out using an Instrutherm UV - 1000A spectrophotometer with an operating range of 320 to 1100 nm, as shown in Equation 1:

$$\text{Chl } a \; (\text{mg/L}) = \frac{[A(663\text{-}750).12.63 \; \textit{Extraction volume} \; (\text{mL}) \, . \, 1000]}{\textit{Vol. filtrate} \; (\text{mL})} \quad\text{..........} \quad (01)$$

Where:
A: absorbance at 663 and 750 nm;
12.63: constant;
Extraction volume: volume of methanol used; Filtered volume: volume

of water filtered.

The procedure for analysing PT was carried out according to the Vanadomolibdic Photocolorimeter method, using the AT100 Alfakit benchtop multiparameter equipment, which follows the methodology proposed by the Standard Methods for the Examination of Water and Wastewater, 18th edition **(Photo 2)**.

The reaction between the phosphate and the reagents causes a blue colour in the sample, with the colour varying according to concentration, using a 10 ml sample of water preserved with H_2SO_4 (Sulphuric Acid).

Photo 2 - Photocolorimeter at-100 PB series E005901 from ALFAKIT LTDA.

Source: The author himself.

The turbidity analysis was carried out according to Esteves (1998) and Tundisi (2008), using a HANNA bench turbidimeter (HI88703), which measures turbidity by comparing the scattering of a light beam passing through the water sample to be analysed with a standard suspension sample. The greater the scattering, the higher the turbidity.

To determine turbidity, the samples were collected in 500 ml bottles and stored in a Styrofoam box with ice until the analyses were carried out in the laboratory using a HANNA Bench Turbidimeter (HI88703) (Photo 4).

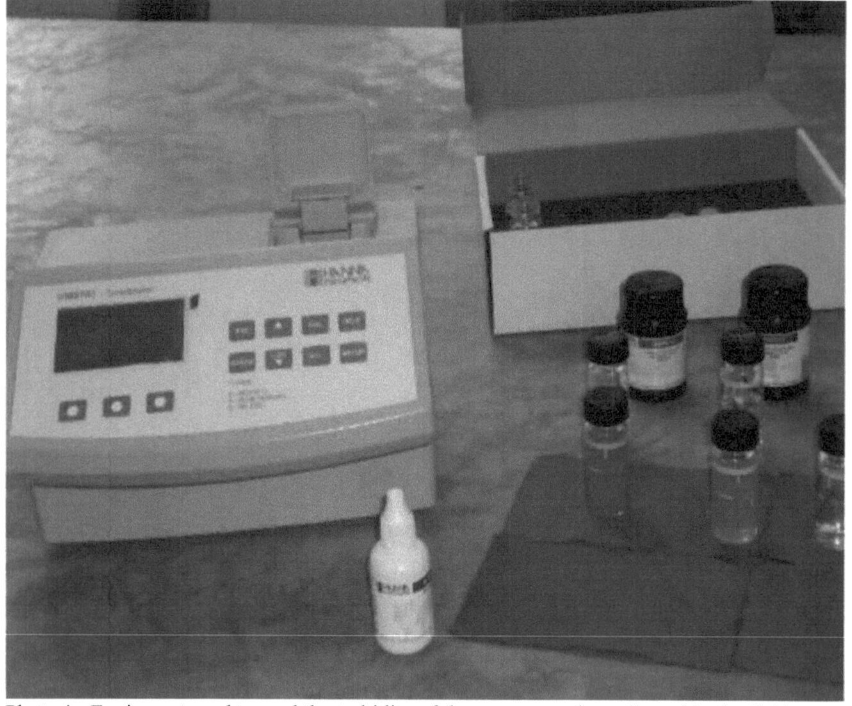

Photo 4 - Equipment used to read the turbidity of the water samples collected in the field.
Source: The author himself.

The turbidity of water is a measure of its ability to scatter radiation.

Quantitatively, this parameter can be expressed in terms of a scattering coefficient or some empirical unit, such as nephelometric turbidity (UNT) (ESTEVES, 1998).

The turbidity analysis was carried out according to Esteves (1998) and Tundisi (2008), using a bench turbidimeter which measures turbidity by comparing the scattering of a light beam passing through the water sample to be analysed with a standard suspension sample. The greater the scattering, the higher the turbidity.

Turbidity was determined using a turbidity meter, which measures the scattering of light produced by the presence of colloidal or suspended particles, indicating the presence of solid materials such as clay, silt and sand and organic materials such as humus or inorganic materials such as oxides. The values are expressed in Nephelometric Turbidity Units (NTU).

The TDS and OD analyses were carried out on site using the Oakton PCD650 multi-parameter equipment. Following the methods described in APHA, the equipment is calibrated in the laboratory before the probe is inserted into the water, after which the TDS and OD analysis standard is selected and the device is left to take the reading (Photo 6A and B).

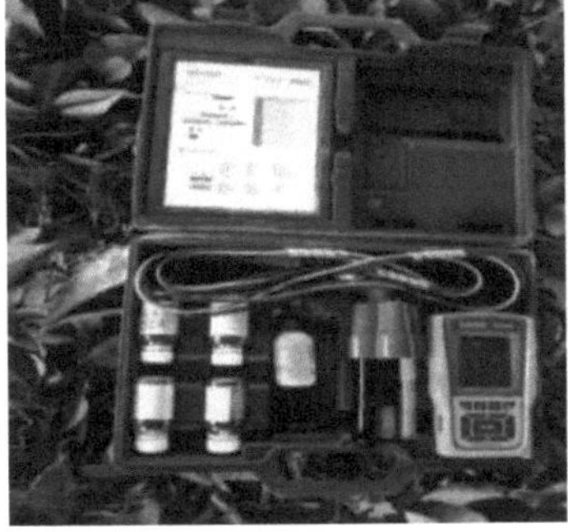 

Photo 6A Multiparameter equipment

Photo 6B Using the Oakton PCD650 device *on site*

Source: The author himself.

TDS is the measure of the concentration of all cations, anions and salts resulting from the combination of cations and anions that are dissolved in water and suspended materials (EMBRAPA, 2004).

1.1.1 **Water Body Framework**

The quality standards and **classification of water bodies were defined in** accordance with CONAMA Resolution No. 357/2005, as described in (Table 2).

According to Branco and Rocha (1977, p. 105), "water quality needs to be assessed, controlled and standardised in order to contain anything that is **foreign to its nature".**

Table 1 - Quality standards for freshwater classes 1, 2, 3 and 4, according to CONAMA Resolution 357/2005.

Parameters	Maximum value Class 1	Maximum value Class 2	Maximum value Class 3	Maximum value Class 4
Chlorophyll "a"	10 µg/L	30 µg/L	60 µg/L	60 µg/L
Phosphorus (P)	0.025 mg/L	0.030 mg/L	0.075 mg/L	0.075 mg/L
Oxygen Dissolved - DO	Not less than 6 **mg/L**	Not less than 5 mg/L	Not less than 4 mg/L	Not less than 2 mg/L
Temperature water	**No standard**	No standard	No standard	No standard
Total solids dissolved - TDS	**500 mg/L**	500 mg/L	500 mg/L	500 mg/L
Turbidity	**Up to 40 UNT**	Up to 100 UNT	Up to 100 UNT	Up to 100 UNT

Source: Adapted from CONAMA 357/05.

1.1.2 Statistical Analyses

The basic statistical data, such as minimum, average, maximum, coefficient of variation and standard deviation, were analysed in accordance with ANDRIOTTI (2003).

For this study, Pearson's correlation was applied, which is a measure of dependence between two variables, used to measure the degree of association between them. Once the correlations (r) had been determined, they were classified qualitatively according to Table 3, and graphs were drawn up with the correlations considered to be strong above.

Table 2 - Qualitative assessment of the degree of correlation between the variables analysed

R	Correlation
0	Nil
0-0,3	Weak
0,3 - 0,6	Regular
0,6 - 0,9	Strong
0,9 - 1	Very Strong
1	Full or Perfect

Source: Callegari-Jacques (2008), Authors' production (2018).

CHAPTER 2

Results and Discussions

2.1 Water Body Framework and Reservoir Compartmentalisation

The classification of the body of water was carried out for each parameter in accordance with CONAMA resolution 357/2005, as described in the methodological procedures.

2.2 Water temperature (Temp)

According to CETESB (2006), water temperature plays an important role in the control of aquatic species and can be considered one of the most important characteristics of the aquatic environment. The water temperature in the reservoir for the rainy season (25/01/13 and 26/01/14) varied between 26° and 29.4°, and for the less rainy season (30/07/13 and 03/08/14), between 21° and 25° (Graph 1).

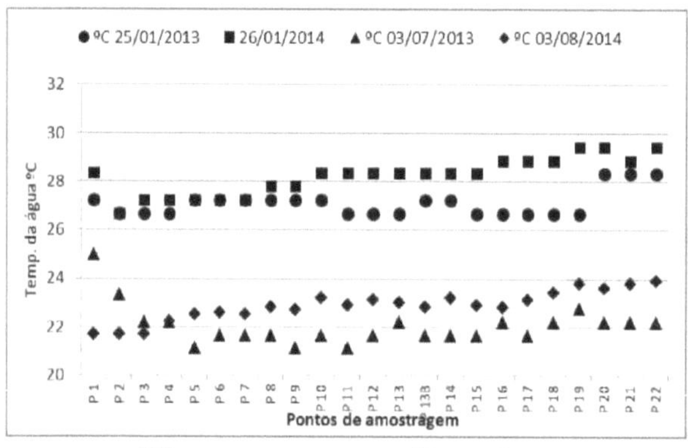

Graph 1 - Water temperature (Temp)

Source: Nogueira. P.F. (2014)

Table 3 - Descriptive statistics for water temperature.

Date	25/01/2013	26/01/2014	03/07/2013	03/08/2014
Minimum	26.6	26.6	21.1	21.7
Maximum	28.3	29.4	25	23.9
Average	27.5	28.0	23	22.8
Standard Deviation	0.56	0.80	0.83	0.64
Coefficient of Variation	2.03	2.86	3.60	2.79

Analysing Graph 1 shows a certain variability in the temperature of the water in the reservoir, and it can be seen that the seasons have an influence on this parameter. The lowest temperatures were found in winter and autumn and the highest in summer and spring, showing that in warm periods the bodies of water tend to gain heat due to the behaviour of the atmosphere, a fact that can be explained by the greater incidence of solar radiation on the bodies of water, while in winter the polar anticyclone over the central-western region causes the air temperature to drop and, consequently, the bodies of water (ROCHA et al., 2014; LIMA; MARIANO, 2014).

2.3 Dissolved Oxygen (DO)

The DO values are shown in Graph 2, demonstrating that for the July fields of 30/07/14, 26/01/14 and 03/08/14, there was little variation in the data and they are within the norms of CONAMA resolution 357 /2005 for class one, where dissolved oxygen cannot be less than 6 mg/L. For the data obtained in the field on 25/01/13, there were oscillations that classified the body of water between classes one and four, as 10 points were in class one, one point in class two, seven points in class three and four points in class four. The lowest OD values were found in the river and transition environment and the highest in the lake environment.

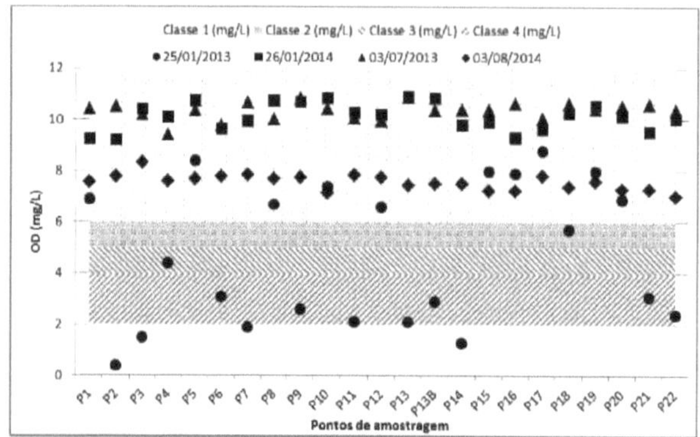

Graph 2 - (DO) Dissolved Oxygen (mg/L)

Source: Nogueira. P.F. (2014)

Table 4 - Descriptive statistics for OD.

Date	25/01/2013	26/01/2014	03/07/2013	03/08/2014
Minimum	0.4	9.22	9.44	7.01
Maximum	8.8	10.88	10.93	8.3
Average	4.74	10.13	10.36	7.57
D.P	2.76	0.53	0.34	0.29
Coefficient of Variation	58.23	5.24	3.26	3.88

Photo 6 - Itaguaçu waterfalls (inside the Itaguaçu district)

Source: Wachholz, (2012).

These fluctuations in DO values (Graph 2) for the samples collected on 25/01/13 may be associated with the rainfall that was occurring during the collection period, possibly carrying organic matter and sediment into the reservoir, altering its equilibrium profile. A fact similar to what happened with the data from Foz do Rio Claro for the collection on 25/01/13 was verified by Piasentin et. al. (2009) in a study carried out in the Tanque Grande Reservoir in Guarulhos (SP), which found a variation in the concentration of DO between January and March 2008, which was lower than class 1 between August 2007 and January 2008, verifying that most of these changes occurred during the rainy season, when organic matter is carried into the reservoir.

Esteves (2011) emphasises the importance of DO in tropical regions, particularly in new reservoirs, where DO is the most critical factor from the point of view of water quality, since in the first few years after filling, the decomposition of the vegetation "flooded" by the lake waters causes very low concentrations of dissolved oxygen in the reservoir body, a fact that can be associated with what happens to the Foz do Rio Claro reservoir, since the vegetation was not removed before flooding the area, and is therefore in a state of decomposition.

2.4 Total dissolved solids (TDS)

According to EMBRAPA (2004), TDS is a measure of the concentration of all cations, anions and salts resulting from the combination of cations and anions dissolved in water and suspended materials. As can be seen in Graph 3, the values obtained for TDS showed little variability for the July 2013 data. In general terms, the values obtained for the 4 campaigns are within the standards of CONAMA resolution 357/2005, which is up to 500 mg/L for class 1 bodies of water.

The TDS varied for all the analyses carried out from 10.6 to 19.9 (mg/L), with the highest values obtained in the two fields carried out during the rainy season (25/01/13 and 26/01/14). The values obtained for the Foz do Rio Claro HPP reservoir are close to the data found by Waterloo et al. (2003) for the Dona Francisca HPP reservoir of 19 to 21 mg/L, which are considered low and do not affect water quality in terms of this variable.

Aquatic ecosystems are greatly influenced by terrestrial ecosystems, so the higher TDS values for the rainy season may be due to solids being carried into the reservoir. This is mainly due to the large area of pasture along the reservoir and the trampling of cattle, which means that during the rainy season there is a greater amount of solids being carried into the reservoir.

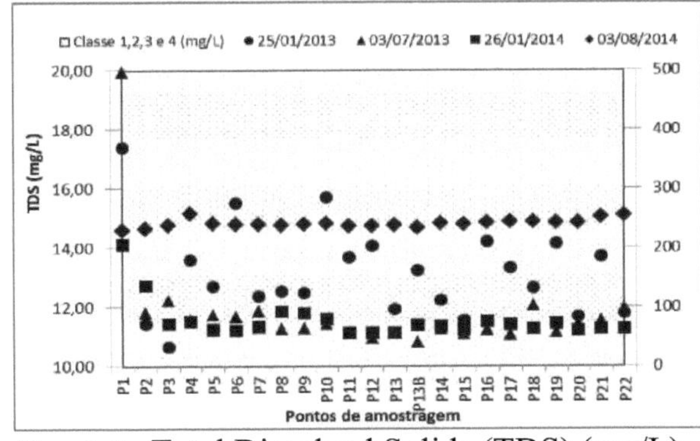

Graph 3 - Total Dissolved Solids (TDS) (mg/L)

Source: Nogueira. P.F. (2014)

Table 5 - Descriptive statistics for TDS.

Date	25/01/2013	26/01/2014	03/07/2010	03/08/2014
Minimum	10.6	10.8	11.1 6	14.
Maximum	17.4	19.9	14.1 1	15.
Average	13.1	11.8	11.5 8	14.
D.P	1.57	1.82	0.66	0.1
Coefficient of Variation (%)	11.93	15.38	5.74 8	0.8

2.5 Turbidity (TURB)

As can be seen in (graph 4), the turbidity for 30/07/2013, 26/01/2014 and 03/08/2014 are within the standard values required by CONAMA resolution 357/2005 for bodies of water that fall into class 1. In the field of 25/01/13 the values obtained were higher, which can be explained by the occurrence of rain during this field and the carrying of sediment by the tributaries, especially point 17.

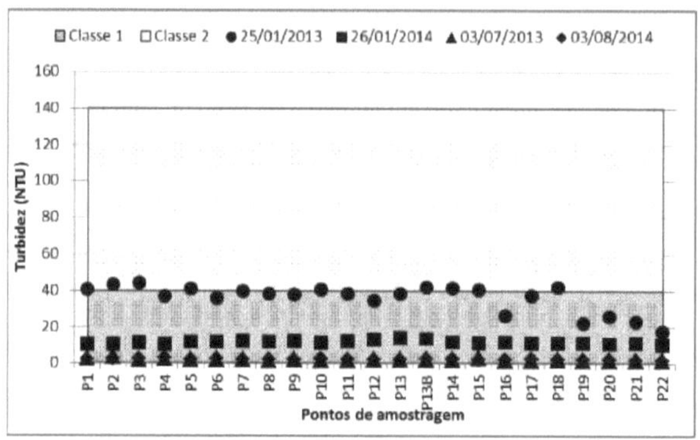

Graph 4 - Turbidity (UNT)
Source: Nogueira. P.F. (2014)

Table 6 - Descriptive statistics for turbidity.

Date	25/01/2013	26/01/2014	03/07/2013	03/08/2014
Minimum	18	10.4	1.8	1.8
Maximum	44.4	14.2	3.8	3.3
Average	36.2	11.9	2.4	2.5
D.P	7.44	1.00	0.47	0.29
Coefficient of Variation (%)	20.52	8.38	20.03	11.30

When comparing the values obtained for the Foz do Rio Claro HPP with those found by Braga (2012) for the Barra dos Coqueiros Reservoir upstream of Foz do Rio Claro, it can be seen that the values are lower than Braga's results, which were between 51 and 72 UNT during the rainy season, when there is more sediment being carried into the reservoir due to the lack of vegetation on the banks, but they are close in relation to the period with low rainfall, which ranged from 0 to 3.7 UNT.

When comparing the data obtained for the Foz do Rio Claro HPP with the data from Santos (2013), the highest results obtained were (22.7 UNT) during the period with the highest rainfall, while the opposite occurred during the period with low rainfall (6.1 UNT), showing that high turbidity values have implications for both the aquatic ecosystem, where it hinders the penetration of the sun's rays, disfavouring photosynthesis and reducing the concentration of dissolved oxygen in the water and other aggravating factors, as well as for treatment and public supply.

According to Richter (2005), water with high concentrations of turbidity and dissolved solids increases treatment costs, reduces the useful life of filters in treatment plants and increases the prices paid by consumers in general. According to Embrapa (2004), turbidity is influenced by the presence of suspended solids, which in high quantities make the water cloudier. The main sources of turbidity are clay, sand, organic waste, mineral material, debris and plankton.

Photo 7 - Compartmentalisation of the reservoir

Sector	Photo	
Rio	A)	 The highest turbidity values occurred in the upstream sector of the lake (river zone). This high turbidity is due to the erosive processes of the banks, the swirling process removing the bottom particles due to the speed of the water flowing over the Itaguaçu Falls and also the discharge of effluents from the urban area.

Transition B)	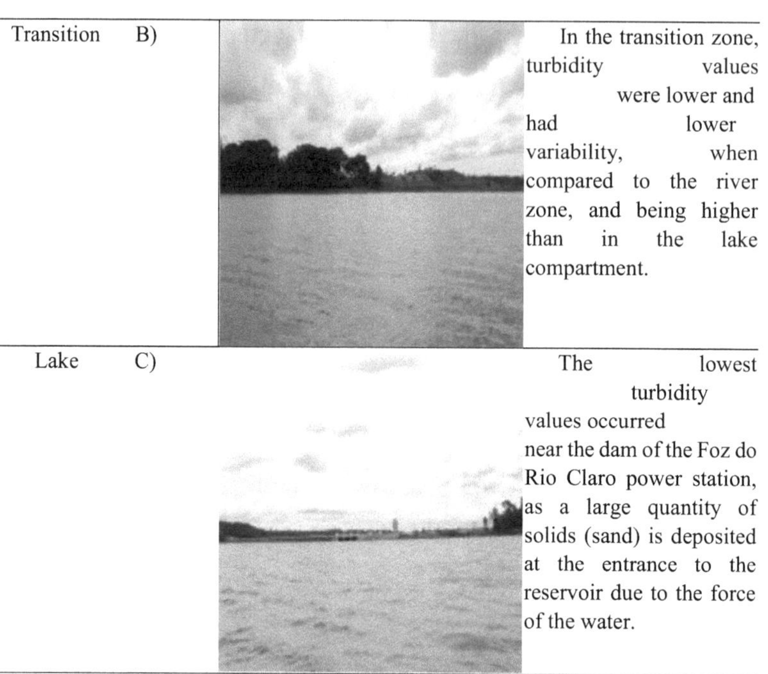	In the transition zone, turbidity values were lower and had lower variability, when compared to the river zone, and being higher than in the lake compartment.
Lake C)		The lowest turbidity values occurred near the dam of the Foz do Rio Claro power station, as a large quantity of solids (sand) is deposited at the entrance to the reservoir due to the force of the water.

Source: Author

2.6Chlorophyll   (CHL)

With regard to CONAMA resolution 357/2005, the results obtained for CHL are **within the standard values for bodies of water that fall into class 1**, with the exception of points 2, 9 and 12, which would fall into class 2 for the fieldwork carried out in January 2013.

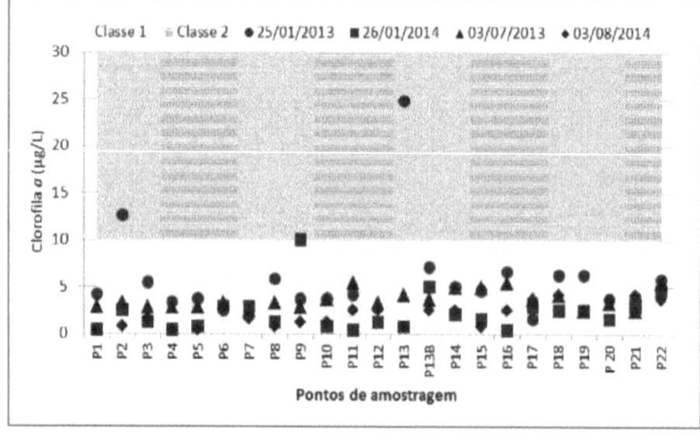

Graph 5 - CONAMA CHL (µg/L)
Source: Nogueira. P.F. (2014)

Table 7- Descriptive statistics for Chlorophyll a.

Date	25/01/2013	26/01/2014	03/07/2013	03/08/2014
Minimum	1,3	0,42	2,53 2	0,4
Maximum	24,8	10,10	5,47 1	4,2
Average	5,6	2,29	3,72 1	2,1
D.P	4,79	2,13	0,98 8	1,2
Coefficient of Variation (%)	84,90	93,30	26,31	60,60

The high value detected for point 2 could be associated with the influence of domestic effluents from the Itaguaçu district that are discharged into the river. Regarding the value found for point 9, we could associate it with the decomposition of vegetation that was not removed from the lake (photos 8), which after 2 years submerged began to decompose. At point 12, the planting of sugar cane could possibly be a factor, since in the previous fields this area was occupied by livestock.

Photo 8: Decaying vegetation near the collection point 9

Source: Nogueira. P.F. (2014)

According to the values obtained for sampling points 2, 9 and 12, and according to Wetzel (1983), in aquatic environments with concentrations above 10 µg/L of CHL, the environments can be considered eutrophicated.

When relating the values of DO (Graph 2) and CHL (Graph 5), it can be seen that where the CHL values are above the limit for classification in class 1, the DO is below 2 mg/L and was classified in class 4 on 25/01/13.

The results for CHL in the fields on 30/07/13, 26/01/14 and 03/08/14 are within the standards established for class 1 according to CONAMA resolution 357/2005.

Based on the data, the values obtained for CHL in the 25/01/13 period are higher than for the period with low rainfall. If we compare the results for the period considered to be rainy found by Câmara (2007) in the Armando Ribeiro Gonçalves Dam/RN for the rainy period, which was (199.2 µg/L) and for the dry period, which was (6.18 µg/L), we can see that the higher CHL values may be influenced by the occurrence of rainfall.

2. 7Total Phosphorus   (PT)

Esteves (1998) and Wetzel, (2001) point out that PT can come from natural sources (present in the composition of rocks, carried by rainwater runoff, particulate material present in the atmosphere and resulting from the decomposition of organisms of allochthonous origin) and artificial sources such as domestic and industrial sewage, agricultural fertilisers and particulate material of industrial origin contained in the atmosphere.

The values established for PT in an intermediate environment, with a residence time of between two and 40 days, for class 1 are up to 0.025 mg/L, class 2 is 0.030 mg/L, and class 3 is 0.075 mg/L P. Based on this, the PT values detected for the Foz do Rio Claro reservoir are above the established standards (Graph 6, Table 8 and Appendix 6) and only in the collection of 26/01/14 are the values within the standards established by CONAMA.

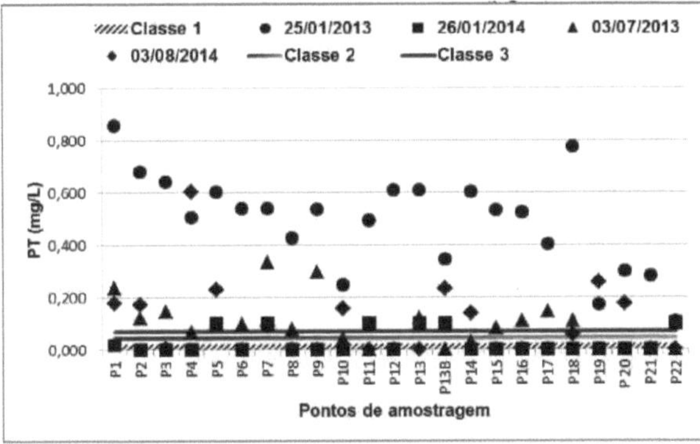

Graph 6 - CONAMA EN (µg/L)
Source: Nogueira. P.F. (2014)

Table 8 - Descriptive statistics for Total Phosphorus (TP).

Date	25/01/ 2 013	26/01/ 2 014	03/07/ 2 013	03/08/2 0 14
Minimum	0,11	0	0	0
Maximum	0,86	0,1	0,33	0,60
Average	0,49	0,03	0,09	0,10
D.P	0,19	0,04	0,09	0,14
Coefficient of Variation (%)	37,48	165,31	101,65	142,90

Another consideration for the high values obtained is the presence of agricultural activities in the reservoir area, where it is common for farmers to use fertilisers containing phosphorus.

Ribeiro Filho et. al. (2011) found PT concentrations of 0.324 mg / L in the Itaipu reservoir and, based on these values, emphasised that land use and occupation are two extremely important factors in water quality studies. Activities such as monoculture and large areas of pasture can increase the concentration of nutrients in the aquatic ecosystem. In the research carried out by Cabral et. al. (2013) in the waters and basin of the Foz do Rio Claro reservoir, it was considered that the type of land use in the study area makes it possible for pesticides to reach the basin's aquatic system, since it is highly productive of grains, especially corn and soya, and is beginning to excel in sugarcane activity encouraged by the government in biodiesel programmes.

Another relevant fact is that the Foz do Rio Claro retains the nutrients and solids that were not left in the upstream reservoirs, noting that this was the last to be **built of the three that feature the run-of-river model.**
An important characteristic to consider for the high (PT) values is that the reservoir has been in operation for four years and may still be in the process of stabilising, and as the area was flooded without cleaning, all the vegetation is decomposing, influencing the results.

According to Von Sperling **(2005), when PT is excessive in a watercourse,** it can lead to the growth of algae, which can cause eutrophication of the **watercourse. Although it does not directly harm human health, high levels of PT** can indicate sources of pollution such as domestic and industrial waste.

2.8 **Pearson Correlation Matrix**

For the period of January 2013, the period with the highest rainfall for the Brazilian Cerrados, there was a correlation considered regular for temperature and transparency, and a strong negative correlation for Total Phosphorus (TP) and Transparency (Table 3), for which the graph was drawn, as the correlation was considered strong (Graph 7).

Table 3 - Correlation for the field of January 2014

	OD	TDS	Turb	Chlorop hyll	EN	Transp
OD	1					
TDS	0,373	1,000				
Turb	-0,177	-0,065	1,000			
5C	-0,115	-0,030	-0,551	1,000		
Chloroph yll	-0,329	-0,305	0,088	-0,213	1,000	
EN	-0,131	0,048	0,700	-0,517	0,153	1,000
Transp	0,039	-0,221	-0,892	0,658	0,022	-0,781	1

Source: Authors' production (2018)

In a study carried out by Araujo (2009) in a reservoir in the Brazilian semi-arid region, it was possible to verify phosphorus concentrations with great variations, in which the minimum values were obtained during periods of low rainfall and the maximum during the rainy season. This corroborates the data found for the research in question, since the period in which there was a strong negative correlation for the PT and transparency parameters is considered to be rainy in the Brazilian Cerrado, when the lowest transparencies and high PT values occur. As can be seen in (graph 7), there was a strong negative correlation for the parameters of PT and transparency, indicating that

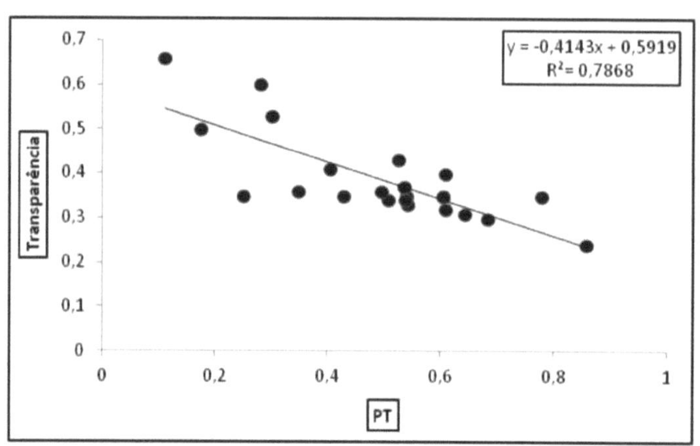

Graph 7 - Pearson Correlation (PT X Transparency) Jan/2013

Source: Authors' production (2018)

Correlation carried out for the field in January 2014 (Table 4)

Table 4 - Correlation for the field of January 2014

Correlation Matrix

	OD	TDS	Turb		Chlorop hyll	EN	Transp
OD	1						
TDS	-0,410	1,000					
Turb	**0,512**	-0,433	1,000				
5C	0,014	-0,209	-0,025	1,000			
Chloroph yll	0,160	-0,067	0,154	0,031	1,000		
EN	0,338	-0,201	0,424	-0,0450	0,030	1,00	
Transp	-0,116	-0,420	-0,223	**0,618**	0,178	0,064	1,000

Source: Authors' production (2018)

According to Siqueira et al (2012), the higher transparency values are due to the monitoring period, in which there was little rainfall on the date of collection, consequently less fluvial contribution, little leaching from adjacent areas and a low rate of suspended material, rather than a relationship with temperature.

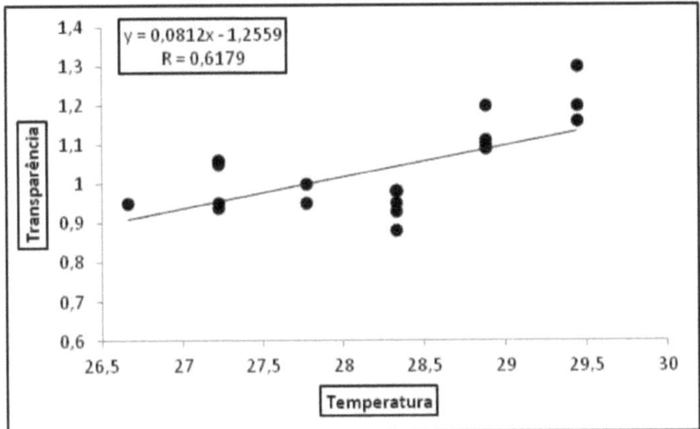

Graph 8 - Pearson Correlation (Temperature X Transparency) Jan/2014
Source: Authors' production (2018)

The correlation for the July 2013 field can be seen in (table 5) where two correlations considered strong were observed, between TDS and °C and a strong negative correlation between Turb and Transparency.

Table 5 - Correlation for the field July 2013

	OD	TDS	Turb	Chlorophyll	P	Transp
				Correlation Matrix		
OD	1					
TDS	0,055	1,000				
Turb	0,076	0,197	1,000			
5C	0,132	**0,799**	0,335	1,000		
Chlorophyll	0,068	-0,210	-0,054	-0,212	1,000	
P	0,307	0,388	0,282	0,155	-0,333	1,000
Transp	0,229	-0,143	**-0,610**	-0,083	0,046	-0,153	1

Source: Authors' production (2018)

There was a strong correlation between TDS and temperature, and for the July 2013 field it was found that as one increases, it influences the other, which can be observed by Lopes (2016) in a study carried out for the Cerrado/Cadunga streams, in which it was also observed that in autumn and winter, the values decrease visibly, with little difference between these seasons, so the TDS values could be higher for the spring and summer seasons.

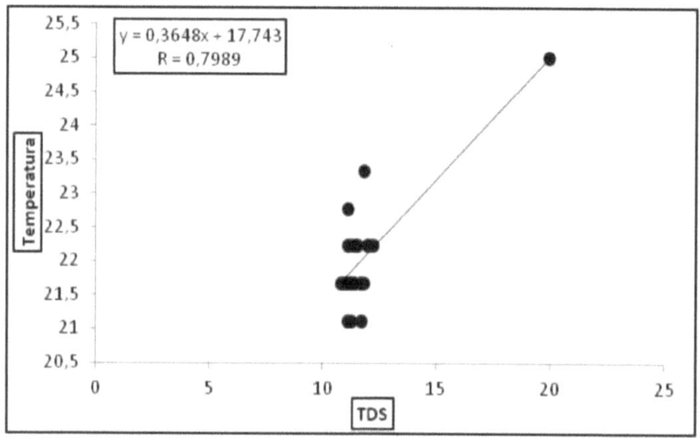

Graph 9 - Pearson Correlation (Temperature X TDS) July/2013

Source: Authors' production (2018)

A strong negative correlation was also found for the Turbidity and transparency parameters, which according to Chagas (2015) turbidity is a water quality parameter that corresponds to the reduction in transparency of the liquid medium, it is promoted by the material in suspension, thus making it difficult for the sun's rays to pass through the water, which was observed for the strong negative correlation obtained for the following parameters, in which it was verified that the lower the transparency the higher the turbidity value.

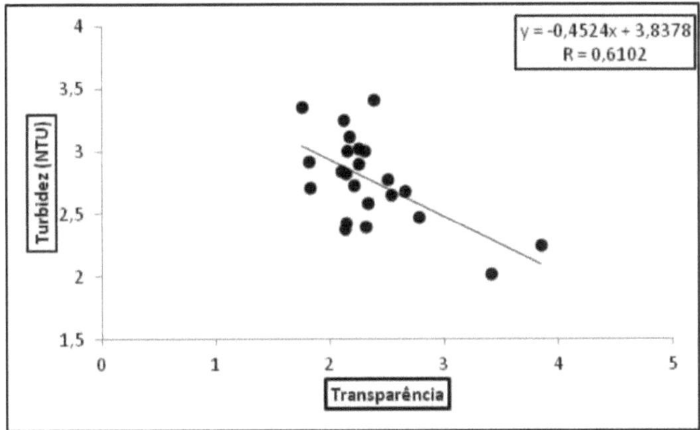

Graph 10 - Pearson Correlation (Turbidity X Transparency) July/2013
Source: Authors' production (2018)

For the August 2014 field season, a period of low rainfall for the Brazilian Cerrado, a strong correlation was found between the **temperature** parameters **and Chlorophyll "a" (Table 6)**

Table 6 - Correlation for the field of August 2014

Correlation Matrix

	OD	TDS	Turb	Temperature	Chlorophyll	P	Transp
OD	1						
TDS	-0,400	1,00					
Turb	0,432	-0,131	1,00				
Temperat	-0,619	0,488	-0,582	1,000			
Chlorophyll	-0,254	0,364	-0,451	**0,694**	1,000		
P	-0,026	0,267	**0,577**	-0,223	-0,348	1,000	
Transp	-0,091	0,004	-0,112	0,004	-0,102	-0,124	1

Source: Authors' production (2018)

The correlation between chlorophyll and temperature was strong, which is justified because one parameter directly influences the other. The higher the temperatures, the higher the chlorophyll values, because algae proliferate more in summer and tend to stabilise in winter when temperatures drop.

Goodin **et. al.** (1993) points out that the monitoring of LCH in lakes and reservoirs is particularly important as it is an indicator of trophic conditions and indirectly of the presence of fertilisers, pesticides and herbicides.

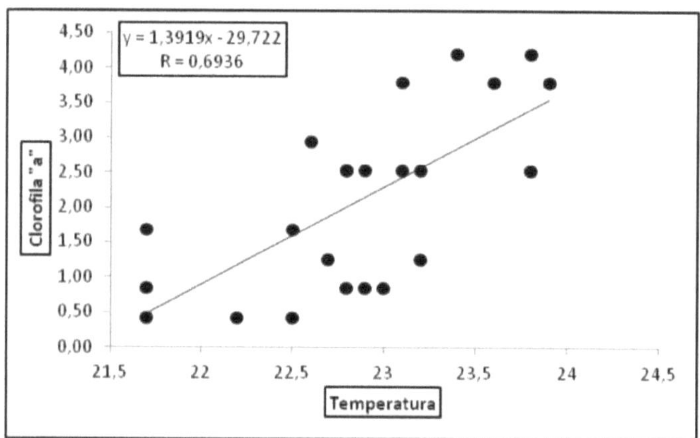

Graph 11 - **Pearson**'s correlation **(Chlorophyll "a"** X Temperature) August/2014

Source: Authors' production (2018)

CHAPTER 3

Conclusions

The classification according to CONAMA resolution 357/2005 showed some parameters above the level established for water class 1 for human consumption.

For the temperature parameter, it was possible to see that the two Cerrado seasons of a hot, rainy period and a cold, dry period influence the oscillations and there are also daily fluctuations according to the times of the analyses carried out.

The results for DO were found to be within the standards set for class 1. Only in January 2013 were some points outside the standards, which could be explained by the rainy season that occurred during the fieldwork.

For TDS, the values in all the fields are within the standards, well below the 500 mg/L set by CONAMA.

Turbidity showed no influence from the periods in which they were carried out, remaining within the established standards. Only in the January 2013 field were some points higher, which can be justified as OD due to the rains that occurred, since with the rains there is a greater movement of sediment into the body of **water.**

The Chlorophyll "a" parameters, in accordance with CONAMA resolution 357, classified the reservoir waters as class 1 and 2 for the first campaigns, but they stabilised, which can be explained by the decomposition of the organic matter present in the reservoir, since the vegetation was not completely removed before the flooding.

The PT values obtained were very high, above the established standards. This parameter may be influenced by the agricultural and grazing practices that occur in the area of the basin in question and also by the deposition of **untreated** sewage **in the body of water, since at the entrance to the reservoir there is the district of** Itaguaçu (GO), from which there is the possibility of sewage being discharged, which may be influencing the high results obtained.

CHAPTER 4

References

ANDRIOTTI, J. L. S. **Fundamentals of statistics and geostatistics.** São Leopoldo: UNISINOS, 2003. 165p.

ARAUJO, F. O. de. **Effects of nutrient enrichment (N and P) under different light conditions on phytoplankton growth in a eutrophic reservoir in semi-arid Brazil.** Dissertation (Master's in Ecology). Federal University of Rio Grande do Norte. 2009

BRAGA, C. C. **Spatial and temporal distribution of suspended solids in the tributaries and reservoir of the Barra dos Coqueiros hydroelectric power plant - GO.** Dissertation (Master's in Geography) Federal University of Goiás. 2012.

BRANCO, S.M; ROCHA, A. A. **Pollution, protection and multiple uses of dams.** São Paulo: Edgard Blucher Ltda. p. 185. 1977

BRAZIL. National Environment Council. Resolution No. 357 of 17 March 2005. **Provides for the classification of bodies of water and environmental guidelines for their classification, as well as establishing the conditions and standards for discharging effluents, and other measures.** Available at: < http://www.mma.gov.br>. Accessed on 12/01/2018.

CABRAL, J. B. P. WACHHOLZ, F. NOGUEIRA, P. F. OLIVEIRA , S. F. BRAGA,C. C. ROCHA, I. R. DA. BARCELOS, A. A. DE. LOPES, R. M. ROCHA, P. R. DA. JUNIOR, V. S. Q. SANTOS, A. K. F. DOS. **Preliminary Characterisation of the Waters of the Foz do Rio Claro HPP - Goiás.** XX Brazilian Symposium on Water Resources. Water. Economic and Socio-environmental Development. 17 - 22 November 2013 - Bento Gonçalves / RS. 2013.

CABRAL, J.B.P. **Analysis of sedimentation and application of forecasting methods to take mitigating measures regarding the siltation process in the Cachoeira Dourada reservoir** - GO/MG. Curitiba - PR. Thesis (Doctorate in Geology, Area of Concentration Environmental Geology) UFPR - Federal University of Paraná - Earth Sciences Sector. 194p. 2006.

CALLEGARI-JACQUES, SINTIA M. **Biostatistics: principles and applications. Artmed,** Porto Alegre. 2008.

CETESB (Environmental Sanitation Technology Company). **Report on the quality of inland waters in the state of São Paulo 2005/CETESB.** São Paulo: CETESB. 488 p. 2006

CHAGAS, D. S. **Relationship between suspended solids concentration and water turbidity measured with an optical backscatter sensor.** Dissertation (Master's in Agricultural Engineering) - Federal University of Recôncavo da Bahia, Cruz das Almas, 2015.

CHORUS, I; BARTRAM, J. **Toxic cyanobacteria in water: a guide to their public health consequences, monitoring and management. London: WHO/UNEP, 1999.**

EMBRAPA. Brazilian Agricultural Research Corporation. **Water quality assessment. - Practical Manual,** EMBRAPA MEIO AMBIENTE, MINISTERIO DA AGRÍCULTURA, PECUÁRIA E ABASTECIMENTO - Brasília- DF, 2004.

ESTEVES, F. de A. **Fundamentos** de **limnologia.** 2ª ed. Rio de Janeiro: Interciência, 548p. 1988.

ESPÍNDOLA, E. L. G., BRIGANTE, J. **Limnologia fluvial: um estudo no rio Mogi-Guaçu.** 1st Ed. São Carlos: RiMa. 278p. 2003.

FERNANDES, L. A. Lithostratigraphic map of the eastern part of the Bauru basin (PR, SP, MG), scale 1:1. 000.000. Boletim Paranaense de Geociências, 55, p. 53-66, 2004.

GOODIN, D. G., et. al. **Analysis of Suspended Solids in Water Using Remotely Sensed High Resolution Derivative Spectra. Photogrammetric Engineering & Remote Sensing,** Vol. 59, No. 4, April, p. 505-510. 1993.

IBGE - Brazilian Institute of Geography and Statistics. Radambrasil Project. Ministry of Mines and Energy. General Secretary. Water Resources Survey. Sheet SE 22 - Goiânia. V31. Geology, **geomorphology, pedology. Rio de Janeiro.** 1983.

KOPPEN, W. **Climatologia con un studio de los climas de la tierra.** Buenos Aires. 320p. 1931.

LATRUBESSE, E. M.; CARVALHO, T. M. **Geomorphology of the State of Goiás and the Federal District. Goiânia, GO**: Secretariat of Industry and Commerce and Superintendence of Geology and Mining of the State of Goiás, 2006.

LIMA, A. M; MARIANO. Z. F. De. **Microclimatic analysis inside and outside semideciduous seasonal forests in the area of the Caçu- GO hydroelectric power station basin.** Revista do Departamento de Geografia - USP, v. 27, p. 67-87. 2014

LOPES, S. M. F. **Influence of land use on water quality in watersheds with different uses, in jataí GO and** Canapolis-MG. Thesis (PhD) - Federal University of Goiás, Institute of Socio-Environmental Studies (IESA). Postgraduate programme in geography, Goiânia, 2016

MACKINNEY, G. **Absorption of light by chlorophyll solutions.** The Journal Biological Chemistry, v. 140, p. 315-322, 1941.

PIASENTIN, A. M; JUNIOR, D. L. S; SAAD, A. R; JUNIOR, A. J. M; RACZKA, M. F. **Water Quality Index (IQA) of the Tanque Grande Reservoir, Guarulhos (SP): Seasonal Analysis and Effects of Land Use and Occupation.** São Paulo, UNESP, **Geosciences**, v. 28, n. 3, p. 305-317, 2009.

QUEIROZ, J. V. S. **Use of Geotechnologies to Characterise the Environmental Fragility of the Foz do Rio Claro HPP Basin (GO)** Monograph presented to the Geography degree course at the Federal University of Goiás - Jataí Campus/ CAJ-UFG. 2014

RIBEIRO FILHO, R. A.A, PETRERE JUNIOR, M.B, C. BENASSI, S.F. D. PEREIRA, J.M.A.A. **Itaipu Reservoir limnology: eutrophication degree and the horizontal distribution of its limnological variables.** Braz. J. Biol, vol. 71, no. 4, p. 889-902. 2011.

ROCHA, H.M, CABRAL, J. B. P., BRAGA, C.C; **Spatio-Temporal Evaluation of the Waters of the Tributaries of the Barra dos Coqueiros Reservoir, Goiás.** Brazilian Journal of Water Resources. , v.19, p.131 - 142, 2014.

SANTOS, Q. R. DOS. FRAGA, M. S. DE. ULIANA, E. M. REIS, A. S. DOS. BARROS, F. M. **Monitoring Water Quality in a Cross-Section of the Catolé River, Itapetinga-BA.** Enciclopédia Biosfera, Centro Científico Conhecer - Goiânia, v.9, N.16; p. 2013.

SIEG. **Goiás State Geographical Information Statistics System.** Available at: <http://www.sieg.go.gov.br>. Accessed on: 10 May 2013.

SIQUEIRA, G. W; APRILE, F; MIGUÉIS, A. M. **Diagnosis of the water quality of the Parauapebas River (Pará - Brazil). Acta Amazônica,** vol. 42(3) 2012

REBOUÇAS, A. DA C.; BRAGA, B.; TUNDISI, J. G. 2002. **Freshwaters in Brazil - Ecological capital, use and conservation.** 2ª Ed. São Paulo: Escrituras Editora.

TUNDISI, J. G. **Reservoirs as complex systems: theories, applications and perspectives for multiple uses.** In: HENRY, R. (Ed.) Reservoir ecology: structure, function and social aspects. Botucatu: FUNDIBIO. p. 19-38. 2007.

TUNDISI, J. G.; MATSUMURA TUNDISI, T. **Limnologia.** São Paulo: Oficina de Textos. 632 p. 2008.

VON SPERLING, M. **Introdução a Qualidade das Águas e ao Tratamento de Esgotos.** 2nd ed. - Belo Horizonte: Department of Sanitary and Environmental Engineering; Federal University of Minas Gerais, 243p. 1996

WETZEL, R.G. **Limnologia.** Calouste Gulbenkian Foundation. 1011 p. 1993.

WETZEL, R. G. **Limnology: Lake and river ecosystems.** California, USA: Academic Press, 2001.

YUNES, J. S; ARAÚJO, E. A. C. **Protocol for analysing chlorophyll-a in water. Rio Grande do Sul:** Cyanobacteria Research Unit of the Federal

University of Rio Grande Foundation. [s/d].

Printed by Books on Demand GmbH, Norderstedt / Germany